SEXUAL HEALTH AWARENESS

What Is

Herpes?

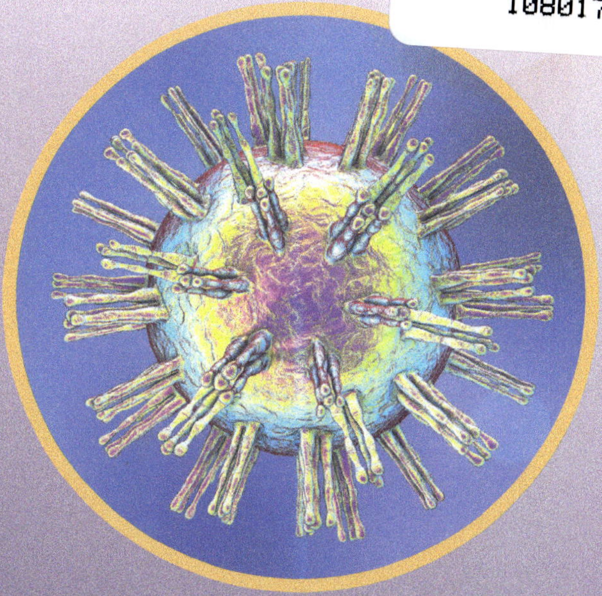

Kathleen Connors

ROSEN PUBLISHING
NEW YORK

Published in 2022 by The Rosen Publishing Group, Inc.
29 East 21st Street, New York, NY 10010

First Edition

Portions of this work were originally authored by Jeri Freedman and published as *Your Sexual Health: Herpes.* All new material in this edition was authored by Kathleen Connors.

Library of Congress Cataloging-in-Publication Data

Names: Connors, Kathleen, author.
Title: What is herpes? / Kathleen Connors.
Description: New York : Rosen Publishing, [2022] | Series: Sexual health awareness | Includes index.
Identifiers: LCCN 2021029088 (print) | LCCN 2021029089 (ebook) | ISBN 9781499472226 (library binding) | ISBN 9781499472219 (paperback) | ISBN 9781499472233 (ebook)
Subjects: LCSH: Herpesvirus diseases--Juvenile literature.
Classification: LCC QR201.H48 C66 2022 (print) | LCC QR201.H48 (ebook) | DDC 614.5/812--dc23
LC record available at https://lccn.loc.gov/2021029088
LC ebook record available at https://lccn.loc.gov/2021029089

Manufactured in the United States of America

Some of the images in this book illustrate individuals who are models. The depictions do not imply actual situations or events.

CPSIA Compliance Information: Batch #CWRYA22. For further information contact Rosen Publishing, New York, New York at 1-800-237-9932.

CONTENTS

INTRODUCTION

There's something inherently scary in the discussion of sexually transmitted diseases, or STDs. Many have very serious health consequences that can change the course of a person's life. The STDs chlamydia and gonorrhea can have complications that can even eliminate a person's ability to have biological children. Some, like human papillomavirus (HPV), can cause cancer. Herpes, while serious, is often an STD that an infected person can live a long and happy life with. There are cases of herpes that are more damaging, but they usually occur in those with a compromised immune system.

Despite that good news, a herpes diagnosis isn't to be taken lightly. It's a lifelong infection that many people have. Perhaps because its long-term effects are often viewed as simply an annoyance rather than a real health condition, many infected people still engage in riskier sexual behaviors. They may not communicate their disease to potential partners if their herpes isn't visible at the time of a sexual encounter. Additionally, some people who engage in unsafe sex may carry herpes without knowing it. Not everyone who has herpes has the telltale blisters and sores. These are just some of the reasons that the statistics about those with herpes are startingly high. The Centers for Disease Control (CDC) reports that one in every six Americans between the ages of 14 and 49 has genital herpes.

Many teens and young adults learn about STDs—sometimes called sexually transmitted infections (STIs) or venereal diseases (VD)—in health classes, but this level of education isn't often presented every year in school. In addition, the facts of how one can contract an STD can get lost in the noise of the media's portrayal of sexual relationships and perceived pressure from peers. Still, STDs are a big problem among young people in the United States. The CDC estimates that half of the new STD and STI diagnoses each year are from those 15 to 24 years old. Yet, TV shows and movies featuring young people engaging in risky sexual behavior with few consequences are still the norm.

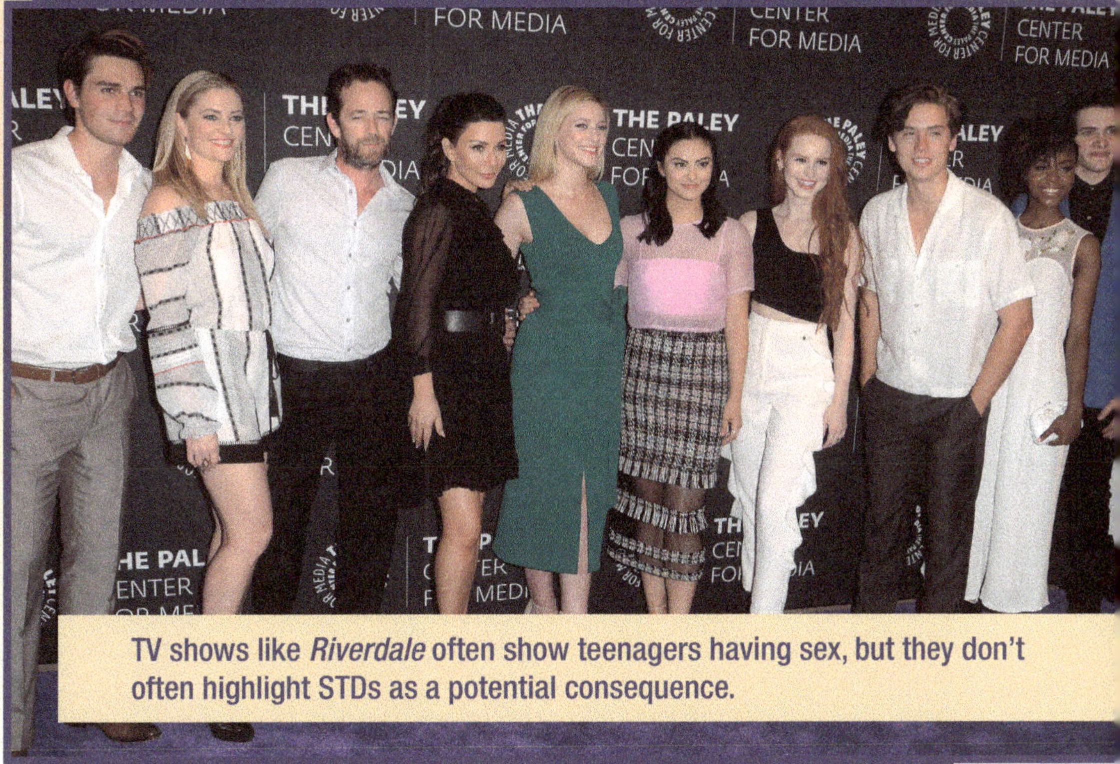

TV shows like *Riverdale* often show teenagers having sex, but they don't often highlight STDs as a potential consequence.

Teens may perceive that this behavior is normal among their peers as well. However, in 2015 the Pew Research Center reported that two-thirds of teens who date said they had not yet had sex.

It can seem daunting and even unnecessary to educate yourself about STDs before you contract one. However, because of how common they are among young people, it's wise to both know what to look for and how to prevent contracting one. If you are sexually active in any way or plan to become sexually active, having tough discussions with potential sexual partners is the first step in protecting yourself from herpes. The next is choosing to abstain from sexual behavior if you are unsure about a person or situation or using proper protections that prevent the transmission of herpes and other STDs. If you do contract herpes, it's important to know how to deal with the physical and emotional effects of it. You also need to consider how to talk to future sexual partners and your medical providers about a herpes diagnosis. Contracting herpes isn't the end of your romantic life, but it is something you'll need to consider in every relationship that follows. Learning about STDs isn't just acquiring knowledge. It's showing respect and care for your health and the health of those you love.

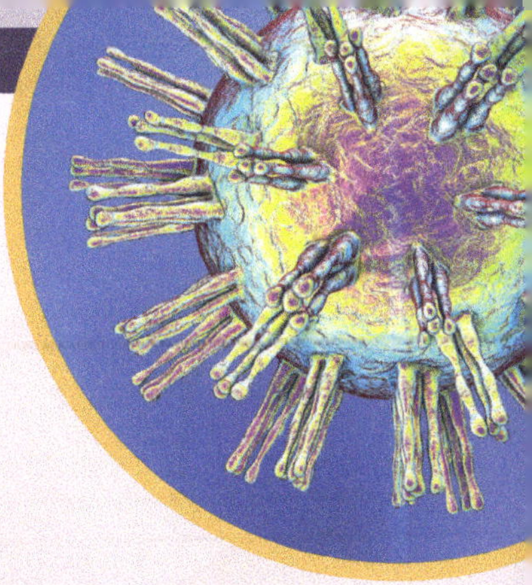

CHAPTER 1

ALL ABOUT THE HERPES VIRUS

Sexually transmitted diseases (STDs) are illnesses one person passes to another when engaging in sexual activity. They can be passed on through sexual intercourse, but that isn't the only way an STD can move from one partner to another. STDs can be passed on through any sexual behavior, including oral and anal sex. That means that STDs can affect any person who has any kind of sex, regardless of age, sex, gender identity, or sexual preference. No one is completely protected from STDs unless they abstain from engaging in sex with a partner.

Like other diseases, STDs can be caused by bacteria, parasites, and viruses. All kinds of herpes, including both oral and genital herpes, are caused by a virus.

STDs like herpes do not discriminate. They simply spread when the conditions are right.

VIRUSES

Understanding how viruses work helps explain how herpes moves from person to person and grows throughout a body it enters.

A virus is a microscopic particle that consists of a core of DNA (deoxyribonucleic acid) surrounded by a protective shell. The shell of the virus contains projections that allow it to attach to a cell and inject its DNA into it. The DNA contains the genetic blueprint for making more virus particles.

To reproduce copies of itself, a virus must take over a cell in a living organism. Cells in the body consist of a central area, called the nucleus, surrounded by a membrane. Viruses like herpes infect people by attaching to the cell membrane and passing into the cell's nucleus, which contains the chromosomes that carry the cell's DNA. The cell has the ability to copy these chromosomes to make more cells like itself. When the virus invades a cell, it injects its DNA into the cell's nucleus. The virus uses the cell's own mechanisms to make copies of the viral DNA instead of the cell's and to encase the new DNA in copies of its own outer coating. When the cell is filled with copies of the virus, it bursts open. The virus spreads to nearby cells, and the process is repeated.

There are different types of herpes viruses. The herpes zoster virus causes chicken pox in children and shingles in adults. Neither of these is an STD. The type of herpes virus that causes the STD herpes is called the herpes simplex virus (HSV). There are two strains of HSV: HSV-1 and HSV-2. HSV-1, or herpes type 1, is what causes cold sores, which are sometimes called fever blisters, around the mouth. HSV-1 can cause genital herpes, but it's HSV-2, or herpes type 2, that causes most cases of genital herpes. HSV-2 generally causes herpes on the genitals, buttocks, anus, and thighs.

The herpes virus most commonly affects the genitals or mouth, although it can also infect other areas, including the eyes. On the fingers, the infection is called whitlows. Occasionally, herpes affects other areas of the skin too.

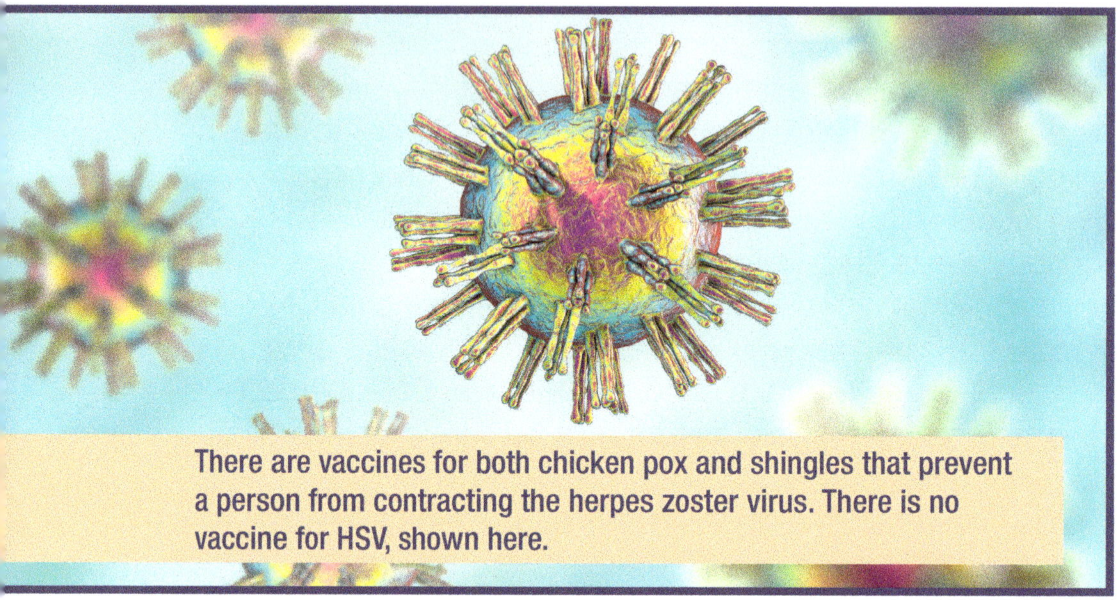

There are vaccines for both chicken pox and shingles that prevent a person from contracting the herpes zoster virus. There is no vaccine for HSV, shown here.

HERPES HISTORY

Herpes infections were common as far back as ancient Greece. The Greek physician Hippocrates (c. 460–c. 370 BCE) described how herpes was spread by contact with infected skin. Ancient Greek scholars dubbed the disease "herpes," from the Greek word *herpein*, meaning to creep or crawl. They gave it this name because of the way the herpes sores spread across skin.

In 1736, French king Louis XIV's physician, John Astuc, reported that herpes was linked with genital contact. The first book on genital herpes was published in 1886 by two French doctors, Charles-Paul Diday and Adrien Doyon.

In 1919, a scientist named A. Lowenstein was able to confirm by experiment that HSV was infectious. In 1925, an American virologist, Ernest Goodpasture, proved that herpes viruses infect the central nervous system as well as the skin, traveling through nerves rather than the bloodstream.

HOW DO YOU GET HERPES?

Herpes is a viral infection of the skin. It's transmitted when the virus is shed from the skin of a person infected with herpes. When herpes is present on the skin or in an active sore, this is called viral shedding. Shedding occurs when the herpes virus becomes active and travels down nerve fibers to the surface of the skin. Viral shedding can take place with or without active symptoms like sores. However, it is more common when symptoms are present. Once it reaches the skin, the virus particles on one person's skin are able to move into the second person's body through any opening in the skin or mucous membranes, such as those in the mouth or vaginal lining.

The herpes virus isn't passed through blood. Therefore, it is not transferable via contact with an infected person's blood or through a blood transfusion. It can be transferred through the saliva of a person presenting with oral herpes, though. People are most likely to be infectious when they are experiencing active symptoms of herpes. However, the virus can sometimes be transferred from the skin when a person is not having symptoms. Once the virus particles are in a person's body, they can infect that person's cells as well as be passed on to other people.

Simply hugging or holding hands with someone won't cause you to contract herpes from them.

MYTHS AND FACTS

MYTH
STDs are only spread by contact with an infected person's blood.

FACT
Herpes is transferred by skin-to-skin contact and is not present in the blood.

MYTH
Only people who have many sexual partners are likely to get STDs like herpes.

FACT
Anyone can catch an STD during sex. This includes herpes.

MYTH
A person knows if they have herpes.

FACT
Many people who have herpes don't know they have it because they have no symptoms or mild symptoms.

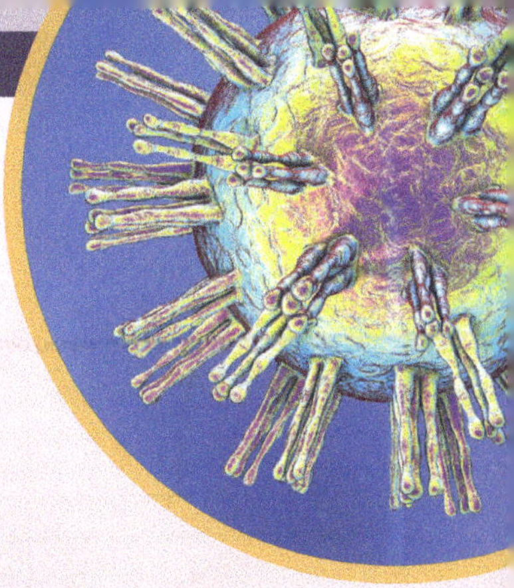

CHAPTER 2

HERPES IN THE BODY

Some people don't experience symptoms of herpes right after contracting it. In fact, they may never have symptoms or not have symptoms until a long while later. This makes it difficult to know when—or possibly from whom—the virus was contracted. It can be an especially troubling thought if a person has had multiple sexual partners: they don't know when they got herpes so they also can't know when they started passing it on.

No two cases of herpes will look exactly the same, but how herpes acts in the body and how the body reacts toward herpes, are the same.

HERPES VS. THE IMMUNE SYSTEM

Herpes can produce a wide range of symptoms. It can produce blisters or sores. In other cases, it just causes a rash. Typically, the visual symptoms are blister-like sores. These sores start as blisters, open up, and then develop a scab. Those infected with herpes may also experience itching, swollen glands, fever, headache, muscle aches, and burning while passing urine. Sometimes, these symptoms present without sores and might be mistaken for the flu. Sores might be mild enough that one thinks they are a pimple or ingrown hair.

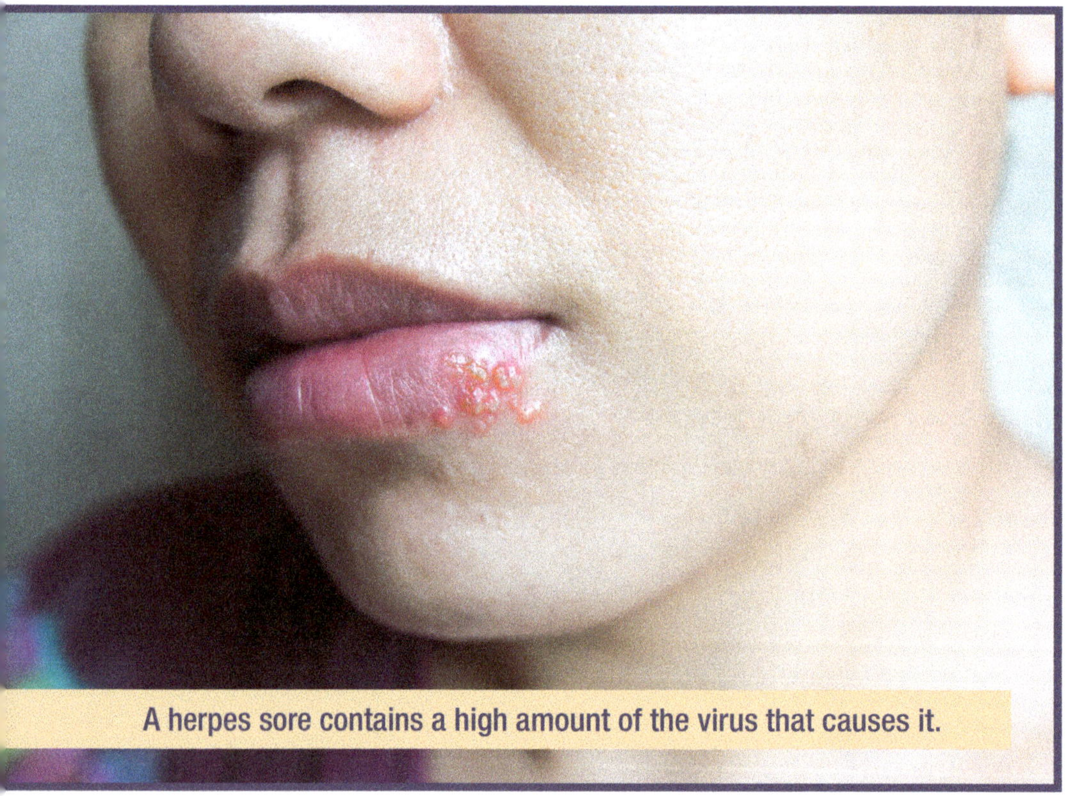

A herpes sore contains a high amount of the virus that causes it.

There are several factors that determine how virulent the herpes virus is. Among these are a gene called ICP47 and one called ICP34.5. These genes interfere with the functioning of elements of the immune system that would otherwise attack the virus.

The body controls an HSV infection through the actions of the immune system. The immune system produces cells that fight viruses and other foreign organisms that invade the body. Some of these cells are produced in bone marrow in the arm and leg bones, and some are produced in the lymph nodes, small glands located throughout the body. The cells produced by the immune system travel through the blood vessels and through lymphatic vessels, which are similar to blood vessels but are filled with a fluid called lymph rather than blood.

Two major types of cells that fight the HSV virus are natural killer (NK) cells and T cells. (The "T" represents "thymus," which is the organ in which these cells usually mature.) NK cells are able to identify the infection in cells. When an NK cell finds an affected cell, it kills the cell, keeping the virus from using it to replicate any more viruses. T cells target the virus particles themselves. The T cells produce antibodies, which are Y-shaped proteins. The antibodies attach, or bind, to the virus, making it impossible for it to engage in any further activity. The first time a person is infected with a virus like herpes, the immune system has to produce antibodies to that particular virus from scratch. Once the body has created antibodies to that virus, they continue to circulate in the body to protect it in the future. That is why the first outbreak of herpes is usually the worst. Subsequent outbreaks are usually shorter and less severe, because the existing antibodies can deactivate the virus more quickly.

Because there's no cure for herpes, the role of the immune system in controlling the herpes virus is especially important. Therefore, keeping the immune system healthy is critical. In most cases, herpes doesn't affect organs other than the skin. However, in those with a weakened immune system, it can affect other parts of the body, including the brain, esophagus, lungs, eyes, liver, and spinal cord. Among those with weakened immune systems are:

- Individuals who have acquired immunodeficiency syndrome (AIDS) or have been infected with the HIV virus, which causes AIDS.
- People who are on medications to suppress the immune system, as a treatment for diseases related to the immune system, such as lupus or Crohn's disease.
- Individuals who take medications to suppress the immune system because of an organ transplant.
- Those who are under stress for long periods of time. Stress causes the body to secrete chemicals called corticosteroids, which help the body function under stress. These chemicals have the side effect of suppressing the immune system.
- People who are very young or very old. In the former, the immune system has yet to fully develop, and in the latter it functions less well.

If you do contract herpes, it's important to make sure any doctor you see knows about it, especially before you undergo any treatment that involves suppressing the immune system.

OUTBREAKS

The theory of latency was developed by Australian microbiologist Frank MacFarlane Burnet in 1939. According to this theory, the herpes virus, when attacked by the immune system, becomes dormant in the roots of nerves. Later the virus can become reactivated, causing another outbreak of herpes symptoms. In this way, the herpes virus differs from many other infectious organisms, which are destroyed by the immune system once they are identified. Herpes sufferers can experience recurring outbreaks of the disease throughout their lives from the same virus they were originally infected with. Burnet lacked the scientific tools to prove his theory. However, it was confirmed by scientists Jack Stevens and Marjorie Cook in 1971.

Not everyone who has an episode of herpes experiences further outbreaks. Those who do often find that subsequent outbreaks are less severe than the first episode because the immune system already has antibodies to the virus and can respond more quickly and effectively. Subsequent outbreaks of herpes are preceded by symptoms such as tingling, itching, burning, or pain near the site where an outbreak will be seen. These serve as a warning that an outbreak is about to occur. People experiencing these warning signs may want to start taking an antiviral medication to reduce the symptoms of an outbreak.

According to the Centers for Disease Control, it's more likely that a person infected with HSV-2 will have another outbreak, or recurrence, of genital herpes than someone infected with HSV-1. Also, those infected with HSV-2 are likely to have an outbreak several times each year, whereas HSV-1 outbreaks usually occur much less frequently, usually only once per year. The occurrence of outbreaks varies from person to person, however. Outbreaks are most likely in the first year or two after infection and tend to decrease over time.

Although it is not known exactly why the virus reactivates at various times, the factors that influence outbreaks are known. Common factors include being run down, having one's period, experiencing prolonged exposure of the affected area to sunlight, catching another STD, drinking large amounts of alcohol, being under stress for an extended period, and experiencing depression. Many of these factors make an outbreak more likely because they negatively affect the immune system, which normally keeps the virus in check.

HEALTHY IMMUNE SYSTEM HABITS

The more stressed out you are about having herpes, the more your body will put out chemicals that suppress your immune system, which makes outbreaks more likely. In turn, the more outbreaks you have, the more stressed you are likely to become. To break this cycle, it is useful to learn some stress-reduction techniques such as meditation, progressive muscle relaxation, or behavioral stress management

GENITAL WARTS

It may be confusing to someone who has noticed a change in their genitals to distinguish between genital herpes or genital warts. Genital warts appear in many of the same places as genital herpes, but they look different. Genital warts may look like small brown, pink, or flesh-colored bumps on your genitals or, when many warts appear close together, may look somewhat like cauliflower. They may be found in the mouth or throat as well, depending on how the warts were contracted. Genital warts are caused by human papillomavirus, or HPV. There are over 40 kinds of HPV that can cause harm in the genitals and reproductive system and some of these cause cancer. Like herpes, HPV and genital warts spread by sexual contact and a person can transmit genital warts without knowing they are infected with HPV.

Because it can be hard to know whether a growth is genital herpes or genital warts, it's important to visit a doctor if you notice these on yourself.

(which teaches you how to control stress-producing behavior and thoughts). There are many books and workshops on these techniques. Your doctor, a local community mental health center, or one of the support organizations for herpes can recommend resources for stress management. Additionally, do things you like to do to reduce stress. Read, learn to knit, or call friends when you are feeling stressed. Simply getting outside for a walk or hike is another great way to quickly reduce stress, whether physiological or emotional.

Spending time in nature is one way to reduce feelings of stress. And reducing stress helps keep your immune system strong.

Other habits that keep your immune system strong include eating fruits and vegetables, exercising, and getting enough sleep. Avoiding or drinking alcohol in moderation is another recommended way to keep the immune system functioning well.

EVERYBODY CAN GET HERPES

While anyone can contract herpes from any kind of sexual contact, there are some populations that are more likely to get herpes. Heterosexual biological women are more likely to have herpes than heterosexual biological men because it's easier for these women to contract herpes from a partner than vice versa. For this same reason, those who are biologically female and have sex with others who are biologically female are at a lower risk of contracting herpes, and those who are biologically male and have sex with others who are biologically male are at a higher risk. No matter how a person identifies or who their sexual partner is, it's important for everyone to remember that herpes and other STDs do not discriminate and safe sex is important in all sexual encounters.

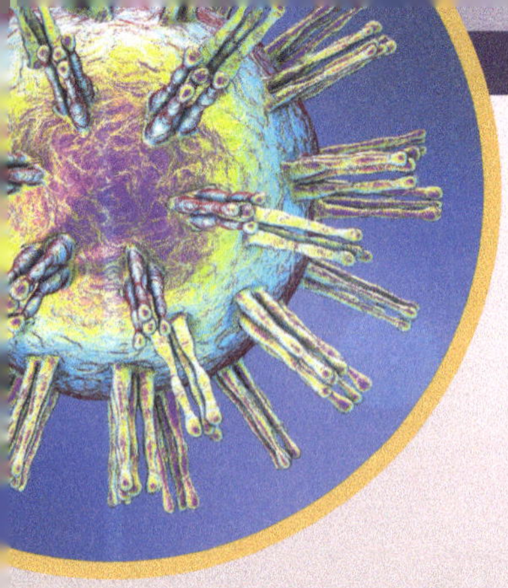

CHAPTER 3

GETTING HELP

If you are concerned you may have herpes at any time, it's important to contact your doctor immediately. Having an accurate diagnosis allows you to get the help you need as soon as possible and begin learning how to manage your symptoms. Additionally, if you have contracted herpes through risky sexual behavior, it's likely your doctor will want you to be checked for other STDs, particularly because some can have grave lasting effects or even cause cancer. Being honest with your health-care team about engaging in any kind of sex is of the utmost importance during these consultations.

DIAGNOSIS

Most healthy children and teens visit the doctor just once a year for their annual physical exam. However, doctors aren't examining patients for herpes at these checkups and wouldn't be able to check for herpes if a person isn't showing symptoms anyway. Even when people visit the gynecologist for a Pap smear, doctors don't test for herpes if no symptoms are present.

For herpes to be identified, a person must have symptoms. If a patient has sores that might be herpes, the doctor will take a swab of the area. This process consists of taking a sample of the fluid from the sores using a cotton swab. The swab is sent to a lab for

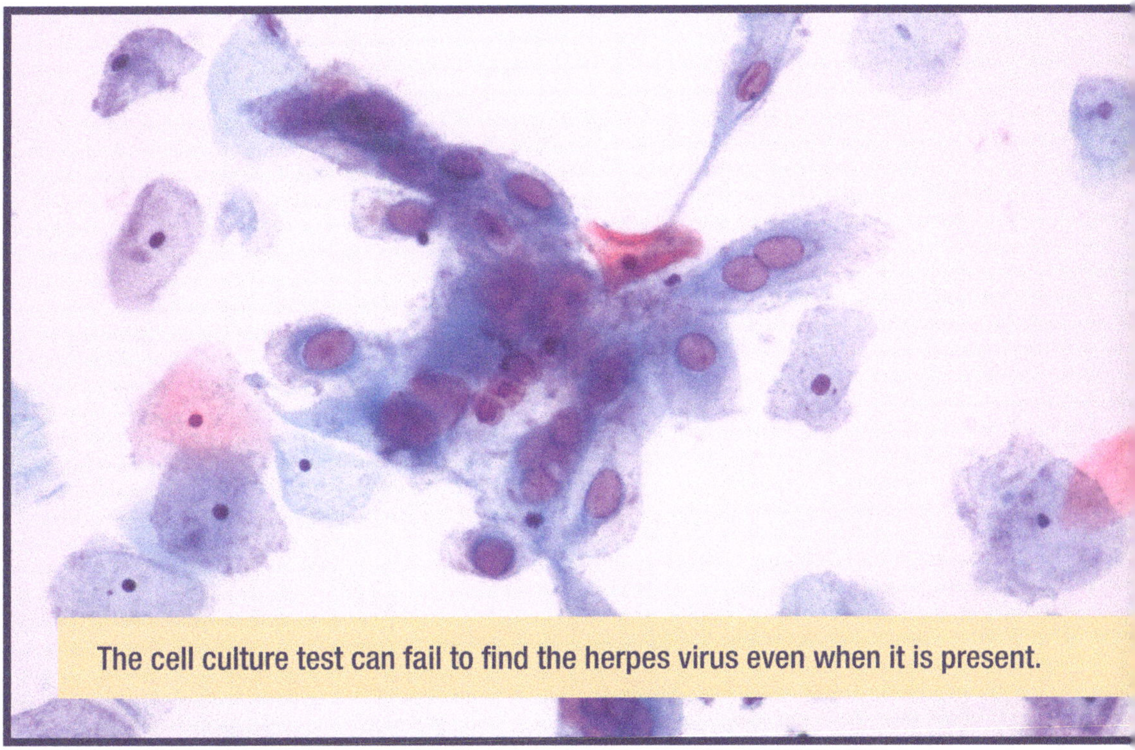

The cell culture test can fail to find the herpes virus even when it is present.

testing, by means of a cell culture. They will put the material on the swab in a medium containing nutrients and see if they can grow cells that show the virus when looked at under the microscope.

There are some issues with the cell culture approach. It can produce a false negative if the infection is very recent or the sores are nearly healed. It can also sometimes produce a false positive. Therefore, if your behavior is such that it's unlikely you've been infected, the test may need to be repeated.

Another test for the herpes virus is the polymerase chain reaction (PCR). In this test, a sample of the viral DNA is compared to the known profile of the DNA of the herpes virus. This test is extremely accurate, but it is more complicated to perform and therefore more expensive than a cell culture.

Immunofluorescence antibody tests can also be used to test for the presence of the herpes virus. This type of test relies on the fact that when foreign particles enter your body, your immune system produces antibodies, which bind to the particles and deactivate them. In an immunofluorescence test, a fluorescent dye is combined with a sample that may contain the herpes virus. Antibodies for HSV are added to the sample, and if the virus is present, the antibodies attach to it. The antibodies are then separated from the rest of the sample and examined under a special light. If viruses are attached to the antibodies, they glow under the light. Because each type of antibody will attach only to a particular virus, the fluorescence shows that the herpes virus is present. Because there are different antibodies for HSV-1 and HSV-2, this test can identify which strain of herpes a person has as well.

Although these tests will identify the fact you have been exposed to herpes, they will not indicate when that occurred or how long you've had it, only that the virus is in your system. Your exposure to herpes could have occurred recently or a long time in the past. There also is no way to compare one person's infection to another. Therefore, there is not an available test to determine from whom someone contracted the virus.

10 GREAT QUESTIONS TO ASK YOUR DOCTOR

1. How long is it likely to take for my symptoms to go away?
2. What can I do to minimize my chances of outbreaks?
3. Should I be tested for other STDs?
4. Is there medication I can take when I have an outbreak to clear it up faster?
5. What can I do to make the symptoms less annoying?
6. Should the person I'm involved with be tested for herpes?
7. When should I return for a checkup?
8. What should I do if my symptoms don't improve?
9. What are the side effects of the treatment you recommend?
10. Do you know of any support groups for people with this problem?

TALKING TO A DOCTOR

Noticing something has changed in your body, especially something that you suspect might be an STD like herpes, can be alarming. Telling a doctor about it may seem intimidating, particularly if you still see a pediatrician that you've known for a long time. Preparing for the conversation can help calm your nerves. Write down your feelings and symptoms and what you suspect may be wrong. Be clear and honest, giving the doctor all relevant facts that they need to know. Make a list of questions to bring with you. Take notes on the doctor's answers to refer to later.

Most states allow teens over 13 to pursue testing for an STD without parental consent. This means your doctor can provide testing and diagnosis without your parents knowing, unless you say they can be told. Telling your parents is a good idea, too, but not everyone's home situation is the same. It's important to seek medical care when it's needed regardless of your relationships with your family.

While your pediatrician or family doctor should be a trusted person for you to speak with, there are other options if you don't have a doctor already or want to speak with a different medical professional. Planned Parenthood and other health clinics provide STD testing and information about safe sex and birth control. These places keep your visits confidential. Your school may also have a health clinic you can visit for testing or information. This may be confidential as well, though it's important to ask if that's the case if confidentiality is something you're looking for.

It's rare, but both HSV-1 and HSV-2 can cause meningitis, or an inflammation of the lining in the brain. For this reason, among others, it's important for your doctor to know if you have herpes.

TREATING HERPES

Herpes cannot be cured. However, it's a disease that can be managed long term. Some treatments help reduce the effects of symptoms when a recurrence happens. There are also medications that can help decrease the likelihood of a recurrence.

Methods of treating herpes began to be developed in the 1960s. These treatments were based on inhibiting the viral DNA so the virus could not replicate (make copies of) itself. Herpes was first officially designated a sexually transmitted disease in 1970. In 1978, Gertrude B. Elion developed the first antiviral prescription medication for herpes. Antiviral medications became available for general use in the 1980s.

The medications used to treat herpes are called antiviral drugs because they are designed to fight viruses. These medications can reduce how long symptoms last and how severe they are. They can also help people with herpes stay symptom-free longer. Among the drugs commonly used to treat genital herpes symptoms are acyclovir (Zovirax), famciclovir (Famvir), and valacyclovir (Valtrex). Most commonly these drugs are taken as pills. Acyclovir is also sometimes given intravenously (through the vein) for severe cases. To be effective, antiviral medications must be taken internally. Applying medications to the skin has little effect.

Treatment is usually given when a person first has signs of an outbreak. Usually, an antiviral medication is given in pill form for seven to ten days. The goal is to improve symptoms or keep them from getting worse.

If the sores have not healed at the end of that time, a second round of medication may be given. Some doctors give patients a prescription for an antiviral drug they can keep on hand to use whenever a flare-up starts. In cases where a person has frequent outbreaks, a doctor may prescribe an antiviral drug to take every day rather than only when a flare-up occurs. This approach suppresses the virus and can significantly reduce or eliminate outbreaks. Suppressive therapy can also reduce the chances of passing on the virus. Nonetheless, those who have been infected should always use condoms when engaging in sexual activity. Luckily, antiviral drugs usually have few side effects.

Be aware that there are many marketers promoting "cures" and "remedies" for chronic (long-term) diseases such as herpes, including special diets and creams. People promoting a "natural" cure sometimes recommend topical substances like coconut oil or petroleum jelly. There is no evidence that these compounds have any effect on the virus (although it's possible that some may have a slight mechanical effect by reducing friction from other tissue rubbing on sores). Also, in most cases, washing with soap and water is just as effective as antibacterial solutions and gels for removing viruses and bacteria from your skin.

Frequently washing your hands with soap and water can help stop you from spreading herpes during an outbreak.

HELP YOURSELF

Keeping your body healthy overall helps keep your immune system strong. Your immune system is your main line of defense against the herpes virus. Keeping your immune system in good shape will allow it to do the best job possible of fighting the virus. To do this, you need to eat a healthy, well-balanced diet. Eat lean protein, vegetables, grains, and healthy fats, while enjoying sweet and fatty foods in moderation. A healthy diet ensures that your body has the nutrients it needs to grow, function, and fight off infections like herpes.

It is also important to get enough sleep. Many young people today are sleep deprived, staying up late when they need to get up early for school and other activities. The CDC recommends teens ages 13 to 18 get 8 to 10 hours of sleep each night. Getting too little sleep affects your mental and physical health and can make your immune system work less effectively. Eating and sleeping right improve your body's strength and energy.

Smoking is another habit that is detrimental to your overall health and should be avoided, especially if you are fighting a chronic disease like herpes. Smoking damages many organs in the body, making your body less able to fight infection.

Avoid excessive use of alcohol and use of recreational, or party, drugs. These can affect the general health of your body and make it less able to fight off infections, including herpes. Using alcohol and recreational drugs can also impair your judgment, making it more likely that you will engage in risky behaviors like unprotected sex.

Most common medications do not affect the immune system. However, some diseases, such as rheumatoid arthritis, lupus, psoriasis, and some forms of cancer, are at times treated with medications that suppress the immune system. Before starting any treatment or medication that might affect your immune system, tell the doctor that you have herpes. They may need to adjust your treatment or give you antiviral medication to prevent an outbreak of the disease.

CAN I HAVE A BABY?

Having herpes doesn't mean you can't have a biological child. Because herpes is not transmitted through the blood, it is not passed from parent to child during pregnancy. If an infection is active at the time a person gives birth, there is a chance the infection could be passed to the baby through skin-to-skin contact. In some cases, if a parent has an active genital infection at the time of delivery, the baby may be delivered by Caesarian section (an incision in the abdomen).

In most cases, parents with herpes can have normal contact with the infant. Because herpes is easily deactivated and removed by soap and water, simply maintaining normal cleanliness is enough. However, it's best to avoid kissing a baby during the first six months of life if active cold sores are present, because an infant's immune system is not fully developed and it may not be able to fight off infection. By six months, the child's immune system should be fully functional and able to fight off the virus by itself.

STAYING SAFE AFTER A HERPES DIAGNOSIS

Once you are diagnosed with herpes, you must know that you can spread it to other people through engaging in any sexual behavior. The chances of spreading herpes to someone else are greatest when you have an active outbreak. Therefore, it's best to avoid sex when you have active symptoms or sores, or until all the sores are healed. Keep in mind that herpes is often spread via oral sex, so you should avoid this activity when you have active cold sores. Using a condom during oral sex as well as intercourse can provide protection. If you or your partner experiences frequent severe outbreaks, taking an antiviral medication on a regular basis may be appropriate.

Exercise helps your body stay in good shape. Having herpes should not keep you from working out, but you do need to be careful if you play sports in which you have lots of skin-to-skin contact with others, such as wrestling. Generally, it's HSV-1 that is transmitted

Competitors and teammates come in very close contact with one another in the sport of wrestling. This could cause the spread of herpes if an infected person isn't careful.

through contact sports. While herpes is only transmitted through mucous membranes and open cuts or rashes, it's still a potential risk to others who may have cuts or skin breakage as well. Your school district may have a policy about notifying others on your team or sports opponents that someone competing on the team has herpes. There may be certain extra cleaning guidelines that need to be followed as well as a protocol you may need to follow during and immediately after an active outbreak of any kind of herpes.

HSV-2 is rarely spread from one person to another other than by having sex. In addition, it is easily removed by the use of soap and water. If you have an active infection and will be in contact with someone whose immune system may not be fully active, such as a baby or an elderly person, you simply need to wash your hands first. Also, it is a good idea not to kiss anyone with a compromised immune system if you have cold sores. At other times, it should be fine to do so.

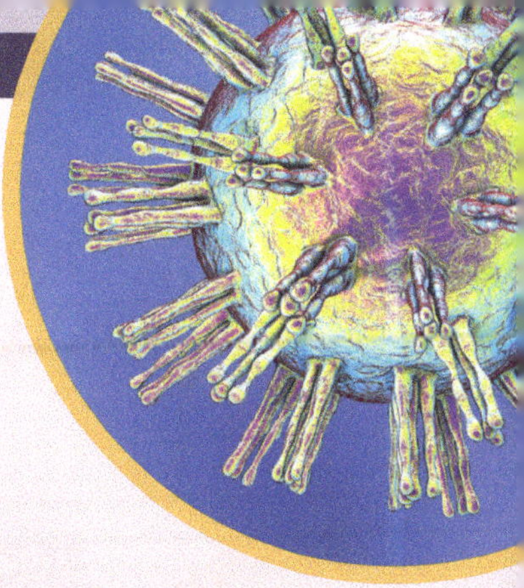

CHAPTER 4

RISKS, PRESSURES, AND PREVENTION

What is your favorite TV show? What music do you like? What celebrities and influencers do you follow on social media? Everything you take in affects how you see the world. And often, the media gives a skewed look at the sexual behavior of teens and young adults through their content. Because of this, many teens and young adults are likely to overestimate the sexual experience of their peers and the amount of sexual experiences they "should" have at a certain age. All of this adds pressure to romantic relationships to become sexual, perhaps before one or both partners are ready. And it might be before partners have discussed serious topics, like the potential for contracting STDs from one another, as well.

Before embarking on a sexual relationship with someone, it's important to know how to protect yourself from contracting an STD like herpes. It's also good to spend time reflecting on your own wants and needs as well as how to maintain good sexual health.

MAITREYI RAMAKRISHNAN

NEVER HAVE I EVER

The Netflix show *Never Have I Ever* is one example of a teen show or movie about a teen who wants to shed a straitlaced reputation and self-image by having sex. However, in these stories, the consequences aren't often highlighted.

THE BIG RISKS

Having unprotected sex is the main factor in the spread of STDs like herpes. "Unprotected" means without a condom. "Sex" in this context means oral and anal sex as well as sexual intercourse. To prevent the spread of STDs, condoms should be used for all kinds of sex. It should be noted that only latex condoms protect against the spread of viruses such as herpes. Even using a condom does not guarantee 100 percent protection against catching herpes, so it is important to take other precautions. You might consider abstaining from sexual behavior altogether, or engage in other sexual behaviors that don't include any mucous membranes across which herpes can cross. If needed, wash after having sex or having any contact with herpes sores.

The next biggest risk factor is having sex with many partners. The more partners you have sex with, the more likely you are to be exposed to the infection. Restricting your sexual relations to one person with whom you have a relationship is safer, though it still carries some risk if you haven't discussed STDs or previous sexual relationships.

Having any kind of sex should not be entered into lightly, and any consequences should be considered ahead of time. That's why using recreational drugs or drinking alcohol is a risk factor for contracting STDs. They affect your judgment and reduce your inhibitions. It's more likely you'll make a decision about sexual behavior you might not have done when sober.

If you do have sex with someone who has herpes, avoid it during active outbreaks when symptoms are present, and have your partner take an antiviral medication to inhibit the virus.

FEELING THE PRESSURE

Do not allow yourself to be talked into having sex before you're ready. How will you know if you're ready? If you feel uncomfortable when the subject comes up, you are probably not ready. If you haven't given any thought to having sex with a person you've been dating, you shouldn't do so until you've thought the matter through. Tell the person you need to think about it. You shouldn't have sex to seem cool, to fit in, to be popular, or because people will think less of you if you don't. You should never agree to have sex with someone because you are embarrassed to say you don't want to. Just say you're not ready, need to think about it, or have decided to wait.

It's absolutely acceptable to tell a person you like them but don't want to have sex yet. Many young people are choosing abstinence, some for religious reasons and some for health reasons. Some simply don't want to risk pregnancy before they prepare themselves for a career or find out what they really want in life. Don't be afraid to say no because the person who asks might reject you. If someone cares about you, they will care about more than having sex.

SEXUAL HEALTH

There are both physical and mental aspects to maintaining good sexual health. Among the mental aspects are developing good communication, goal-

SAYING NO—OR NOT YET

"No" is a complete sentence. However, sometimes simply telling a partner "no" when they ask about sex may not seem like enough, or it might feel too final. Here are some other responses to try:

- I've decided to wait.
- I'm not ready.
- I can't support a child.
- Not until we're married.
- If you loved me, you wouldn't ask.
- School comes before sex.
- It's against my religion/values.
- We're too young.
- I don't feel comfortable.
- I don't want to contract an STD.
- Not everybody's doing it. I'm not.
- We can find other ways to express our love.
- Have you thought about consequences?
- Ask me again in five years.
- I don't want that kind of pressure.
- I said no, and I mean it!

Consent is an important part of engaging in sexual behavior. Consent means all parties need to agree they want to take part in sex at the time it is occurring.

setting, and decision-making skills. These skills are useful for any important decision you make in life, including if and when to have sex and how to do so in a safe way.

Goal setting is the process of deciding what you want in various aspects of your life. Make a list of what you want in terms of personal relationships and other aspects of your life. This process will provide you with the motivation to behave in a manner that will maximize your chances of having what you want in the future. Decision-making skills are tools to help you make choices that are best for you. First, clearly state the decision to be made. Next, research information that can affect whether or not you should do the activity in question. List the pros and cons of each possible course of action, and choose the best option. To a large extent, your decisions as to what is best will be based on your goals.

Communication skills allow you to communicate clearly and comfortably with other people. They are especially important in situations that may be awkward, such as discussing sex. Preparation is one key to good communication. If you don't feel you're ready to have sex or don't feel comfortable asking a potential partner about exposure to diseases, spend some time thinking about what you'll say if you are asked to have sex. Talk it through with a friend so that you're prepared when the subject comes up. Another aspect of communication is talking to others, friends or

Educating yourself about sex, how it affects your health, and the consequences of risky sexual behavior is also part of maintaining good sexual health.

professionals, for advice. Often people feel they must decide things for themselves, but talking to others may provide ideas you haven't thought of. Other people may have dealt with similar situations and developed ways of responding that can be of use to you. They can also provide another perspective of the situation.

From a physical perspective, it is important to make sure you are healthy and to detect any problems as soon as possible. If you become sexually active, have regular checkups by a doctor with whom you have been honest about your sexual experience. They will be able to observe potential problems you may not notice. If you experience any discomfort in the area of your genitals or the lower area of your body, or if you have any unusual discharge, see your doctor immediately. If you have been sexually active and have not been tested for common STDs, you should get these tests. If you have an issue, prompt treatment will increase your chances of addressing it and minimizing discomfort.

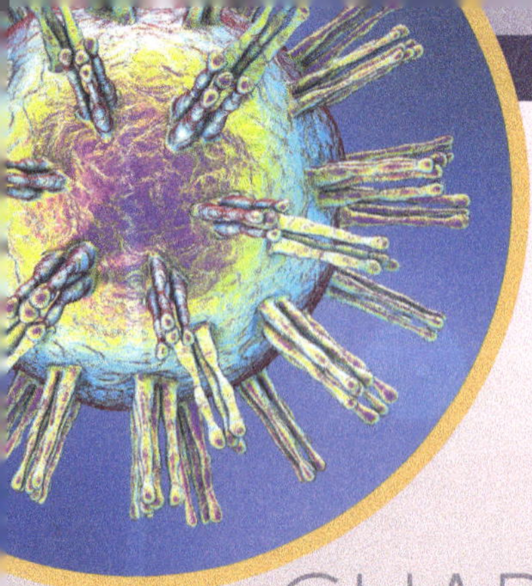

CHAPTER 5

LEARNING TO LIVE WITH HERPES

The effects of herpes are more than just physical. Those who contract it often have an emotional and psychological response as well. They might feel ashamed because of the stigma surrounding STDs. There is also the stress of having to deal with a chronic disease. If you are prone to recurrences, you will have to make sure you have appropriate medication, and you will have to take care not to subject yourself to undue stress or engage in behaviors that are likely to lead to flare-ups. Second, there is the awkwardness of needing to tell potential partners that you have herpes, which can be stressful. However, there are ways to have the tough conversations about being infected with herpes and resources that can help you cope with having herpes.

HOW DO YOU FEEL?

It's not uncommon to feel sadness, anger, and confusion after finding out you have herpes. You may feel angry at the person who passed the disease on to you and angry at yourself for not engaging in safer sex. You may also find yourself withdrawing from other people because you believe that they will reject you or think less of you. Cutting yourself off from other people can have negative effects, leaving you feeling alone and isolated. This can make your situation more traumatic at a time when you are unsure of how to deal with the disease. Feeling stigmatized can lead to self-consciousness and low self-esteem. Support is very important when dealing with a condition like herpes.

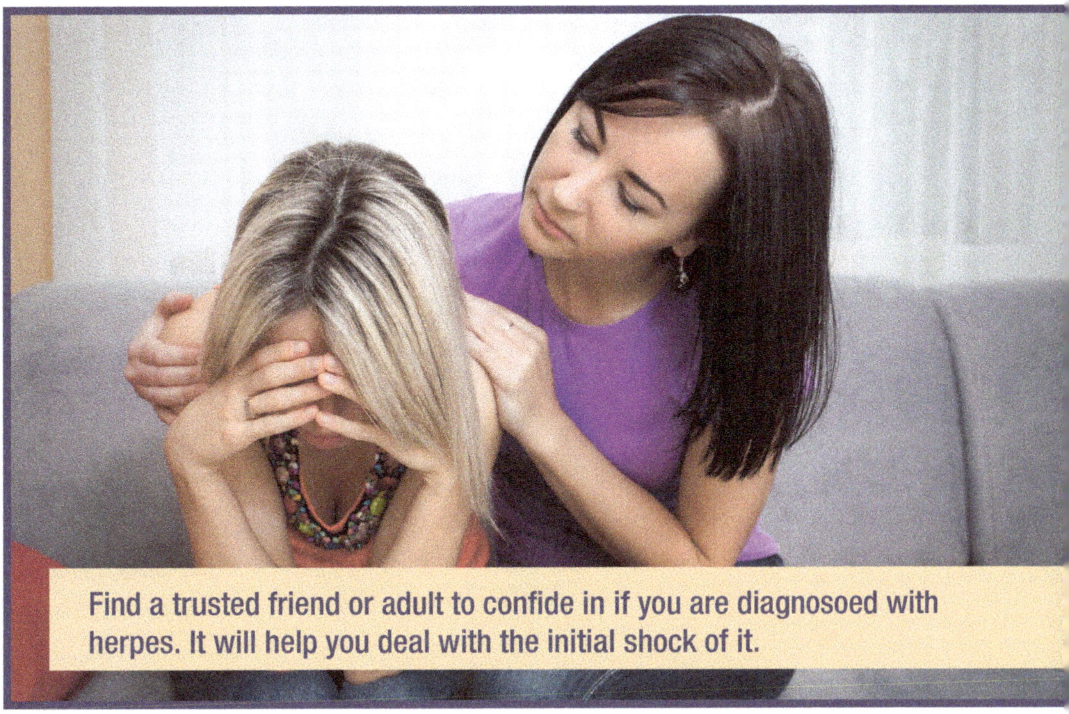

Find a trusted friend or adult to confide in if you are diagnosoed with herpes. It will help you deal with the initial shock of it.

TOUGH CONVERSATIONS ... WITH YOUR PARENTS

Some of the feelings teens and young adults deal with when being diagnosed with herpes might have to do with how they think their parents will react. They might be worried their parents will be disappointed or even angry. They might be afraid to tell their parents at all. However, there are advantages to letting your parents know about having herpes or that you suspect you might have it.

First, they are often in the position to help you get appropriate medical care. They are likely to already have a gynecologist (for those who are biologically female) or urologist (for those who are biologically male) who can help you. Second, your family can be a source of emotional support. Most parents love their children and want to help them. Even if they are upset to hear you have the disease, they will probably want to help you. Their help can provide both practical and emotional support when you are dealing with the problem. This may relieve some of the stress of dealing with it on your own. Also, trying to keep the condition a secret from your family will add to your stress, which is not good for you mentally or physically.

Although in many states it's legal for doctors to treat teenaged patients for sexual issues without telling their parents, your parents may get a bill or a summary of charges from their insurance company if you see your doctor. It's better for them to find out about the issue from you than from a letter in the mail. It shows trust

Depending on your age, you may find that what shocks your parents the most is the fact that you have been sexually active.

and the fact that you are depending on them for help.

Even though it may turn out well, talking to your parents can be a tough conversation. Here are some tips to make it go more smoothly:

Pick a time and place when you can discuss the matter at length without distractions.

Pick the parent you feel most comfortable with and talk to them first. That parent can help you tell the other one once you've talked about the situation and will make the discussion feel less like a confrontation.

Admit you're nervous about talking to them. That will help make the situation feel less awkward.

Tell your parents you need their help with a problem. That will focus the discussion on the problem, rather than on the behavior that led to it. This approach may help turn the discussion into one of working together.

Gather some information about herpes before you tell them you have been exposed. When you tell your parents, they may not know what the implications of having herpes are. You can reassure them that it won't keep you from having a normal life or a family.

Suggest seeing a doctor together, so they have the facts regarding the disease.

Remember that your parents were once teenagers too and not all the choices they made were correct. No matter how upset they are about the situation, taking care of you is still likely to be their major concern.

If for some reason you feel you can't tell your parents, you still need to take care of your health. You could reach out to a school counselor or other trusted adult. Also remember there are resources such as Planned Parenthood and community clinics where you can get tested for STDs and that can recommend resources for you.

TOUGH CONVERSATIONS ... WITH YOUR PARTNER

If you are considering having sex with someone, it's important that you tell them you have herpes. This can be a very difficult conversation. You may be worried you'll be rejected, but that's a risk you need to take for the health of your partner. Don't wait until you are about to have sex—or worse, until after you've had it.

Be prepared to explain what herpes is and that there

It's important to remember that some cases of genital herpes are caused by HSV-1. This means your partner should know if you have an outbreak before engaging in oral sex.

TALKING TO A PARTNER ABOUT STDS

Don't be embarrassed to talk to your romantic partner about STDs like herpes! Even if you aren't having sex now, if you are considering it at all, you should have this conversation. Remember, STDs are medical conditions, and it's up to you to protect your health and your partner's health. Here are some talking points to cover:

- Tell the person you're seeing that it's awkward to talk about the subject.
- Find out about their sexual experience and history and share yours honestly. This will help you both assess the risks of becoming intimate.
- If your partner has had sex before, suggest that you should both get tested for STDs so you can start your new sexual relationship with confidence. It doesn't hurt to have noted a clinic or two where you can be tested and the cost. Sometimes reluctance turns out to be more financial than personal.
- Insist that if you have sex, you should use a condom.

are steps to take to minimize infection. A caring partner will listen as you explain about the disease. Then, it's your turn to listen. They may have fears or questions that you'll need to consider. Doing research together on how to be safe or visiting with a medical professional beforehand is a great way to keep your partner safe and even become closer as a couple. There are safe ways to have sex when you have herpes and a diagnosis should not keep you from becoming intimate with someone in the future.

FIND MORE SUPPORT

When dealing with a chronic disease, turning to your friends and family for support is a great idea. Talking to them about your feelings can help relieve some of the stress of having the disease. Do not feel you have to struggle with herpes alone. Casual contact won't pass herpes to other people, so spend time with your friends. Not only can they provide emotional support, but they can also distract you so you aren't obsessing about the problem. The more you engage in normal social activities, the better you'll feel emotionally.

The second line of support is a professional, such as a psychiatrist, psychologist, or counselor. These experts have the training to help individuals cope with trauma and stress. They are especially useful if having herpes leads you to experience depression. They are also an excellent source for learning coping techniques such as methods of relaxation.

A third source of help is support groups. These groups consist of people who all suffer from the same

It's up to you if you want to tell your friends you have herpes. But it may help relieve stress and help combat the stigma of the disease.

or similar disorders. In a support group, you do not have to fear a negative reaction to the fact that you have a disease. Because the people in the group have the same problem, they know how you feel. Those who have had the problem for a long time can offer practical as well as emotional support. They can share coping techniques that worked for them. In addition to providing emotional support, people in support groups share useful information, such as new treatments.

Support groups can be in-person groups or online. Your doctor or the local hospital or clinic may be able to provide information on support groups that meet in your area. You can also do an online search for "herpes support group [your city or town]." When you are first diagnosed, you may be more comfortable with an online support group that allows you to be anonymous. There may also be Facebook, Reddit, or other social media sites with support group options. When dealing with advice you receive from people over the internet, remember that it is just like talking to people in person. The information you get is the result of their personal experiences and opinions. Although you might want to test their tips, before you follow any health advice, be sure to check with a health professional.

Meeting with others who have herpes in a support group may help you feel less shame as they can show how you can continue to live your life happily.

Finally, there are hotlines, such as the National Herpes Hotline ([919] 361-8488 or the National STI Hotline [800] 227-8922), which can provide you with encouragement and information about symptoms, support groups, and helpful resources.

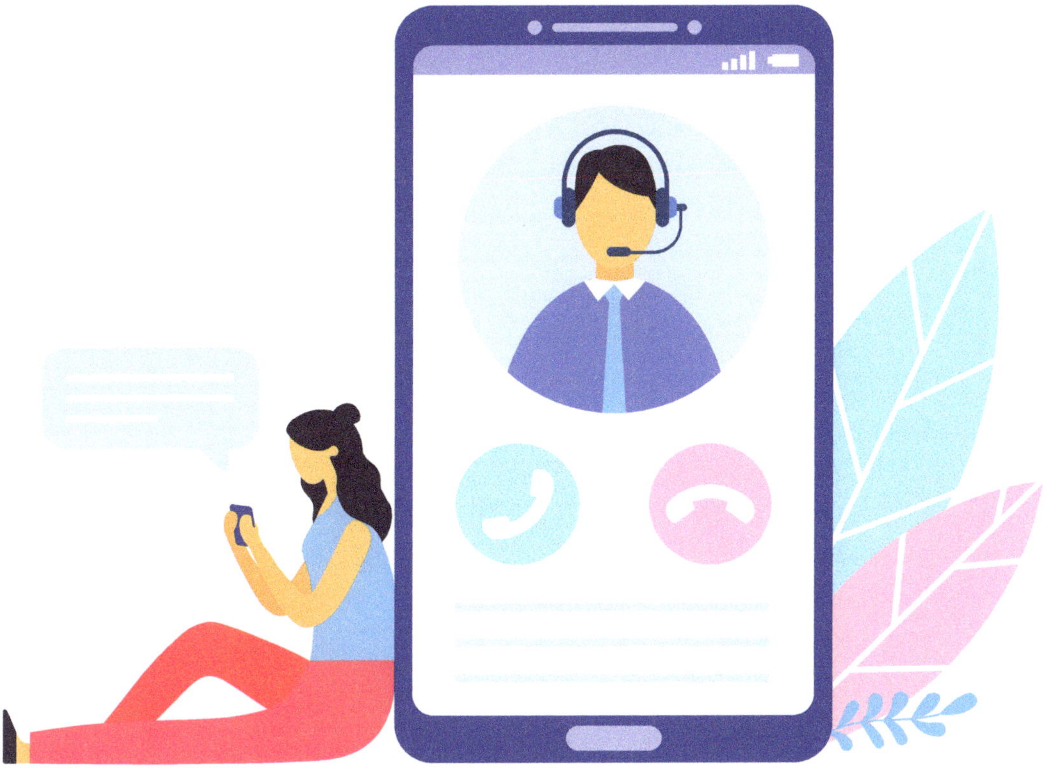

GLOSSARY

abstain To avoid completely.

anonymous Not named or identified.

antiviral medication A medication that keeps viruses from replicating.

bacterium A single-celled organism that can cause infections. The plural form is bacteria.

Caesarian section A process in which a baby is delivered through an incision in the abdomen.

chromosomes The elements in cells made up of DNA, which form the genetic blueprint of an organism.

chronic Going on for a long time.

confidential Secret or private.

corticosteroids Chemicals secreted by the body that prepare it to deal with a threat.

Crohn's disease An ailment in which a person's bowels become inflamed.

deactivate Make inactive.

depression A feeling of intense sadness and helplessness.

DNA (deoxyribonucleic acid) The material that encodes genetic information.

dormant Inactive but alive.

false negative A test result that shows you don't have a disease when you do.

false positive A test result that shows you have a disease when you don't.

gene Part of a chromosome that contains the genetic code for one characteristic of the body.

genital Related to the external reproductive organs.
gland An organ that produces a chemical that affects the body.
inhibit To keep from performing any activity.
inhibition A nervous feeling that stops you from doing or saying something.
intravenous Put into a vein.
latency A period during which a virus remains inactive but is still present in the body.
lupus A disease in which a person's immune system attacks their own body.
mucous membranes The tissue that lines areas such as the mouth and secretes mucus.
nucleus The area in the center of a cell that contains the chromosomes.
organism A living creature.
Pap smear A test done periodically on those who are biologically female in which a small sample of vaginal tissue, usually the uterine cervix, is checked for cancer cells.
perspective Point of view.
profile In genetics, an arrangement of DNA from a particular organism.
replicate To make copies of.
stigma A negative label.
suppress To make less effective.
topical Relating to something put on the skin.

FOR MORE INFORMATION

American Sexual Health Association
P.O. Box 13827
Research Triangle Park, NC 27709
(919) 361-8400
Website: www.ashasexualhealth.org
The ASHA strives to increase education about sexual health, the prevention of STDs, and supporting these cause through public initiatives.

American Sexually Transmitted Diseases Association
PO Box 12665
Research Triangle Park, NC 27709
(919) 861-9399
Website: www.astda.org/
This scholarly organization champions research about STDs and disseminates accurate information about treatment as well as combats the social stigma surrounding the diseases.

Centers for Disease Control (CDC)
National Center for AIDS/HIV, Viral Hepatitis, STD, and TB Prevention
1600 Clifton Road
Atlanta, GA 30329
(800) 232-4636
Website: www.cdc.gov/nchhstp
This part of the CDC works to prevent the spread of STDs and provides data and information to the public.

Education + Training + Research (ETR)
5619 Scotts Valley Drive, Suite 140
Scotts Valley, CA 95066
(800) 620-8884
Website: www.etr.org/
This non-profit organization works to bring education about health, including sexual health, to teachers, workplaces, and young people.

Healthy Teen Network
1501 St. Paul St. Suite 114
Baltimore, MD 21202
(410) 685-0410
Website: www.healthyteennetwork.org/
Dedicated to educating teens about various health issues, including sexual and reproductive health, the Healthy Teen Network provides resources to learn and be proactive about teen health.

Planned Parenthood Federation of America
434 West 33rd Street
New York, NY 10001
(212) 541-7800
Website: www.plannedparenthood.org
Planned Parenthood provides resources for coping with and preventing STDs, as well as other sexual health services such as STD testing and birth control methods.

FOR FURTHER READING

Alderton, Molly et al. *In Case You're Curious.* Jersey City, NJ: Viva Editions, 2019.

Ford, Jeanna Marie Ford. *Understanding Reproductive Health.* Minneapolis, MN: Essential Library, an imprint of Abdo Publishing, 2021.

Hartman, Ashley Strehle. *Teens and STDs.* San Diego, CA: ReferencePoint Press, Inc., 2019.

Honders, Christine. *The Dangers of Sexually Transmitted Diseases.* New York, NY: Lucent Press, 2018.

Mirk, Sarah. *You Do You: Figuring Out Your Body, Dating, and Sexuality.* Minneapolis, MN: Twenty-First Century Books, 2020.

Parrish, Jacqueline. *Coping with Sexually Transmitted Diseases.* New York, NY: Rosen Publishing, 2020.

Quist, Jeremy. *Body and Mind: LGBTQ Heath Issues.* Philadelphia, PA: Mason Crest, 2020.

Samuels, Gary. *The Herpes Survival Guide: Steps to Living Happy and Healthy After Herpes.* Charleston, SC: CreateSpace Independent Publishing, 2012.

Sheen, Barbara. *Teen Guide to Managing Stress and Anxiety.* San Diego, CA: ReferencePoint Press, Inc., 2021.

Young, Frankie. *The Self-Care Kit for Stressed-Out Teens: Healthy Habits and Calming Advice to Help You Stay Positive.* Chinchester, UK: Vie, an imprint of Summersdale, 2021.

INDEX

T

U

V

W

PHOTO CREDITS

Cover and chapter heads, p. 10 Kateryna Kon/Shutterstock.com; series background Lu Vesper; p. 5 Kathy Hutchins/Shutterstock.com; p. 8 Prostock-studio/Shutterstock.com; p. 11 santypan/Shutterstock.com; p. 14 Prasan Maksaen/Shutterstock.com; p. 19 BlurryMe/Shutterstock.com; p. 21 BalanceFormCreative/Shutterstock.com; p. 22 Komsan Loonprom/Shutterstock.com; p. 27 Monkey Business Images/Shutterstock.com; p. 30 New Africa/Shutterstock.com; p. 33 JoeSAPhotos/Shutterstock.com; p. 36 (Maitreyi Ramakrishnan) AdMedia/MediaPunch/AP Images; (Never Have I Ever) https://en.wikipedia.org/wiki/Never_Have_I_Ever_(TV_series)#/media/File:Never_Have_I_Ever_Title_Card.png; p. 38 karelnoppe/Shutterstock.com; p. 41 SpeedKingz/Shutterstock.com; p. 42 Diego Cervo/Shutterstock.com; p. 45 Kaspars Grinvalds/Shutterstock.com; p. 47 LightField Studios/Shutterstock.com; p. 49 Tirachard Kumtanom/Shutterstock.com; p. 50 domnitsky/Shutterstock.com; p. 52 exebiche/Shutterstock.com; p. 54 AS photostudio/Shutterstock.com; p. 55 Tartila/Shutterstock.com.

Designer: Michael Flynn
Editor: Kristen Nelson